GOOD JUNK

GOOD JUNK

by Judith A. Enderle   illustrated by Gail Gibbons

ELSEVIER/NELSON BOOKS
New York

Library of Congress Catalog Card Number 80-26315   ISBN 0-525-66720-2
Published in the United States by Elsevier-Dutton Publishing Co., Inc.,
2 Park Avenue, New York, N.Y. 10016.
Printed in the U.S.A.   First edition
10 9 8 7 6 5 4 3 2 1

My mother just doesn't appreciate good junk!

Once I found a great piece of rope. Of course, it was a little bit dirty. But that was only because it had been in the street for a long time.

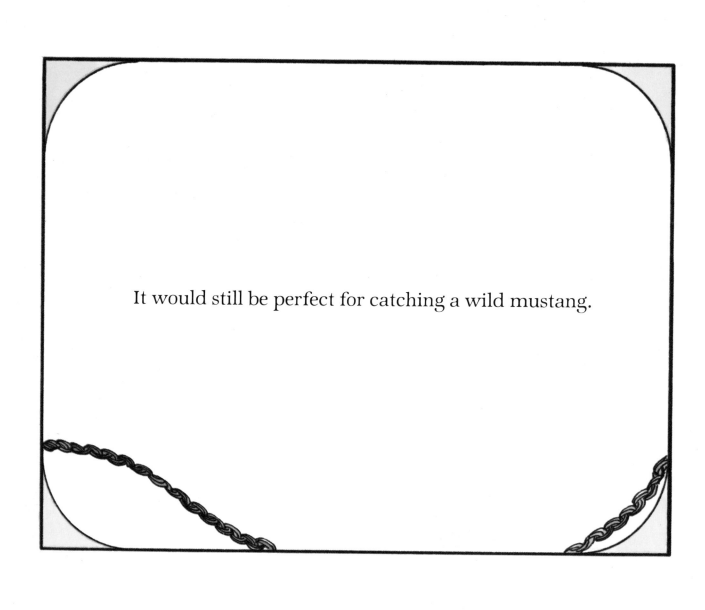

It would still be perfect for catching a wild mustang.

Or if a lion escaped from the zoo, I would need that rope to tie him up so he couldn't eat my mother. Someday she might be sorry she threw that good rope away. So far, though, she always seems glad to get rid of stuff.

Last summer Mr. Bertram tossed out a big box and a bunch of boards. Tommy Curtis and I spent all day hauling them to my house. We were going to build a neat clubhouse. But—you guessed it!

My mother said, "Kirby, take that junk back to Mr. Bertram."

Tommy asked his mother if we could bring it to *his* yard. But Mrs. Curtis said the same thing my mother said. Mothers! I guess they're all alike.

Freddy Ferguson found a real shell in the field by his house. Was he lucky! It still had a sea animal living inside it. He took it home and put it in his treasure drawer. But a week later you could hear *his* Mom all the way down the block. She said his junk smelled!

Once my friend Penny Potter gave me a perfectly good pen—except it didn't have any ink. I used it to write invisible words and secret codes. No one could read them. But my mother didn't think that was anything special. She said, "A pen without ink is just junk." Then she threw it into the wastebasket.

Penny Potter better not want her good junk pen back.

The other day I found a nice, smooth, black tire. I spent a lot of time rolling it home. I was really glad the bigger kids didn't see it first. That's the kind of junk they like to collect. It would have made a super swing. I was going to save it until the tree out back gets big enough. Or I thought I might use it for the go-cart I'm going to build. I'd only have to find three more tires just like that one.

But my mother said, "Get rid of that old thing." My mother definitely does not appreciate good junk!

When we grow up, my friends and I are going to be real junk collectors, just like old Sammy, the guy who comes around in the clunky red truck. We're going to get a truck of our own. On the door it will say GOOD JUNK in big, shiny, black letters. Then, we'll go all over town and collect as much good junk as we can find. We'll have to find a safe place to keep it, though. We won't want any little kids getting into it. Maybe by then our mothers will like junk and we can keep it in our yards.

Right now, my Mom doesn't seem to like anything I find. But wait 'til she sees what I found today on my way home from school.

She can't say he's junk because he's alive. His name is Homer—'cause he needs a good home. He won't clutter up the yard and he won't take up any room in the garage, either. He can stay in my room, right by my bed. I already made a special box for him.

"Hey, Mom! Come and see what I found."